Guidelines for the Tutor of Mathematics

An Aid for Student Tutors, Paraprofessionals, and Teacher Assistants

By

Henry S. Kepner, Jr.
Department of Curriculum and Instruction
University of Wisconsin—Milwaukee
Milwaukee, Wisconsin

and

David R. Johnson
Chairman, Mathematics Department
Nicolet High School
Milwaukee (Glendale), Wisconsin

National Council of Teachers of Mathematics

1906 Association Drive, Reston, Virginia 22091-1593

Fifth printing 1994

Library of Congress Cataloging in Publication Data:
Kepner, Henry S
Guidelines for the tutor of mathematics.
Bibliography: p.
1. Mathematics—Study and teaching. I. Johnson, David R., joint author. II. National Council of Teachers of Mathematics. III. Title.
QA 11.K49 510'.7'1 77-6426
ISBN 0-87353-030-6

The publications of the National Council of Teachers of Mathematics present a variety of viewpoints. The views expressed or implied in this publication, unless otherwise noted, should not be interpreted as official positions of the Council.

Printed in the United States of America

Table of Contents

Introduction

This pamphlet grew out of a need recognized by tutors, teacher assistants, and paraprofessionals for a set of guidelines and suggestions for effective tutoring in mathematics. There is a fast-growing involvement of individuals as tutors in several settings: the mathematics resource center, the classroom, and private tutoring in the student's home.

The prospective tutor, although frequently successful in mathematics, usually lacks teaching skills. This pamphlet presents guidelines from a variety of sources: the authors' experiences, research on tutoring, and direct work with student tutors.

The pamphlet is short enough to be read in a single sitting. However, *the ideas and suggestions will be most valuable when specific sections are reread at appropriate times in the tutor's experience*. The guidelines are not commandments, but they do provide suggestions that have proved valuable to numerous tutors and teachers in the past.

When there are several tutors in a school, this booklet can serve as the basis for tutor training and discussion sessions. The art of tutoring can be developed more fully when tutors and teachers can share problems and suggested techniques in discussions on a regular basis.

The Challenge: Communication

You probably picked up this pamphlet because you have been asked by a parent, student, or teacher to tutor someone in mathematics. If you accept that invitation, you will be undertaking a very worthwhile challenge that can provide satisfaction for you and success for the person you assist. If you are a student, closeness in age and interests and the opportunity to work with only one or a few learners will provide advantages for you that a teacher often lacks.

Students seek tutoring for many reasons. Becky may have been absent from class. Don may not have the appropriate mathematics background for the course in which he is presently enrolled. Al may not be able to learn at the pace set by others in his class. Kevin may get a tutor because he is not learning math effectively the way it is presented—for example, in a lecture setting or in an open classroom. Denise may need individual help because she is reluctant to ask questions in a class setting. There are other reasons, too—an extended illness, a lack of motivation, or a family vacation, for example.

Whatever the reason, you must develop a trusting relationship between you and the student you will tutor. The student must believe that you have a sincere desire to help him or her learn specific mathematical concepts. The student must also feel that you see her or him as a worthy human being, capable of mastering these concepts.

This guide addresses specific facets of the tutoring program in different chapters for easy reference. It is suggested that you first read the booklet in its entirety and then study each chapter in depth. This will give you, the tutor, time to plan how to incorporate these ideas in your tutoring program. The remaining parts of the booklet are organized as follows. Chapter 2 deals with developing the effective communication that is necessary for effective tutoring; chapter 3 is concerned with the necessary steps in planning each session with your student; chapter 4 deals with the actual tutoring session and the activities for both the student and the tutor; chapter 5 deals primarily with the evaluation of the student's progress and the role that you, the tutor, take in this evaluation process. The annotated Bibliography contains a list of resources you may refer to concerning teaching strategies, activities, and mathematical games appropriate for use in the tutoring sessions.

Tutoring is not easy. You must have patience, you must plan carefully, and you must learn to let your student learn by attempting the work himself or herself. Your diagnosis can help the student over the rough spots. You must be honest with your student, but you must demonstrate confidence that she or he can accomplish the task.

But for all its difficulties and hard work, you yourself have much to gain from tutoring. First, like most tutors, you will probably find a personal satisfaction in helping someone else succeed. Second, you will discover along with other beginning teachers that "you never know a topic well until you have tried to teach it." You can learn a concept on a superficial level, but only when you try to explain it carefully to another will you be compelled to explore the ideas in depth, finding their power, applications, and limitations. Finally, your success as a tutor adds greatly to the total learning experiences in the school. Frequently a teacher is not available to provide intensive help to a student who needs it. Your tutoring contribution helps a school meet its responsibility to provide productive alternative learning situations.

Tutoring can be exciting and rewarding. It is the hope of the authors that the guidelines set forth here will contribute to the success of your tutoring experience.

The First Sessions

THE first few sessions in working with any student will be the most crucial. In these sessions you must develop a working relationship with your student—let us call him Brad. With his teacher's help you should seek to identify Brad's mathematical competencies, to diagnose his learning difficulties and strengths, and to help him clarify his own goals for the tutoring experience as they are related to the course in which he is enrolled.

As the tutor, you must work to establish a positive setting for each session. This requires demonstrating a sincere interest in helping Brad improve his mathematical performance. You must show by your actions that you don't regard him as "stupid," "dumb," or inferior because he has difficulty with some mathematical exercises.

It is understandable that both you and Brad will be concerned and nervous at the first meeting. He may have difficulty admitting that he has problems with mathematics, and you may be nervous about whether he will accept you. Approach this meeting confident that you and Brad will be able to work together.

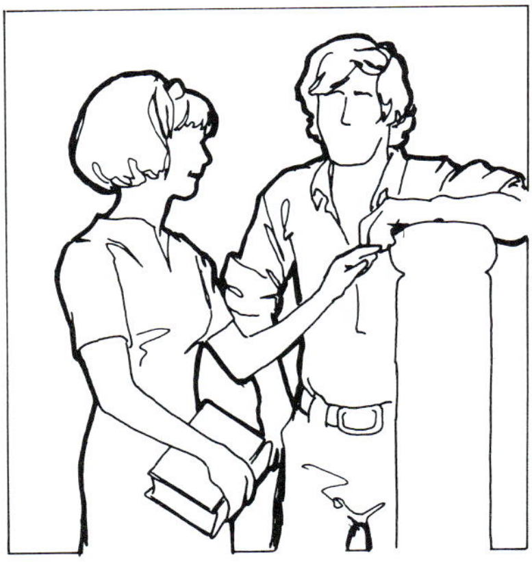

THE TUTORING LOCATION

To begin with, arrange a location where there is minimal distraction. In school, a table in the library or resource center, an unused classroom, or a conference room are good locations. Working in the back of an occupied classroom or in the hall is not recommended, but this may be necessary. Use a table or movable desk so that you and Brad can sit side by side. Don't sit across from him. Sitting next to him will convey a desire to work with him, not to control him. Seek a table in a work area away from distractions and free of interruptions. Arrange the session, whether it is at home or at school, so he won't be inhibited by friends or family hearing him make mistakes.

STARTING THE FIRST SESSION

The first session might well start with an informal discussion about interests, friends, activities, and so on. You may want to identify common experiences that can make the sessions less threatening to your pupil. However you start, be friendly and let your desire to help *show*!

AN INFORMAL SETTING

Try to create an informal setting; don't copy the more rigid teacher-pupil classroom relationship. Brad probably needs help because he is having difficulties in a more formal learning situation. You can observe his problems, both mathematical and emotional, and alter your lesson accordingly.

Brad's respect for you, his tutor, is based on your conduct—both verbal and nonverbal. You must show concern and patience through your actions (especially your facial expressions) as well as your words.

LEARN THE STUDENT'S GOALS

To work effectively with Brad, you must know what goals he has established for the tutoring experience. Is he interested in learning math for his own use or to pass a test or course? Don't push to change his goal, but plan to work within its limits. It will be helpful to know if he sought help on his own or was forced to get tutoring by his parents or a teacher. You can't change these factors, but you can avoid pushing him too hard if he is already under severe pressure from others.

By knowing his reason for being in a tutoring session, you can more easily deal with his feelings of frustration. As you have probably observed, individuals may say something unkind to you in a moment of frustration. Often such a remark may really be intended for themselves or others. Don't be too hasty to get upset yourself.

With regard to mathematics, Brad may have goals that are unrealistic or too broad. He may expect to "learn to do mathematics" or to "master algebra" in these sessions. You should help restate these goals in more specific terms. Instead of such broad goals, suggest that he expect to learn to "do long division" or to "solve linear equations with integer coefficients." State objectives that can be accomplished in a reasonable time and that will be noticeable to everyone involved.

IDENTIFY THE STUDENT'S PERFORMANCE LEVEL

The aim of mathematics tutoring is to take students from where they are to where they should be. To do this, you must determine what they can do in terms of course objectives.

During the first sessions with Brad, you should determine his performance level. Arrange a conference with his teacher, and ask for an evaluation of his performance. With the teacher, develop a list of key objectives for him to achieve during the tutoring experience. For the purposes of this booklet, consider this definition of the word *objective*: *a statement that describes what the student should be able to do following your instruction and his practice.*

This statement describes what is expected of the learner in such a way that it is possible for both the tutor and the learner to determine whether or not the objective has been achieved. For example, suppose a student is expected to learn to factor an expression of the form $(a^2 - b^2)$ with integer coefficients. A poor statement of the objective might be the following: "A student should be able to do algebra." This objective is too vague. It is not possible to determine whether this objective has been met without a definition of "algebra." A clear statement of an objective can be reached through a discussion with the teacher, an examination of the text, and a check of supplementary materials.

Some typical objectives follow. The student should be able—

1. to divide whole numbers when the divisor has one or two digits;
2. to use scale drawings to determine the actual measurements of objects;
3. to add fractions where the denominators have no common factors;
4. to interpret information from charts;
5. to solve equations of the form $ax + b = c$, where a, b, and c are integers;
6. to recognize congruent geometric figures.

If possible, examine samples of Brad's recent work—homework, quizzes, tests. Check with the teaching staff to determine if test questions are available to measure his competence on the list of objectives. If they are available, use appropriate questions in your first few sessions to see what he can do. If not, you should select a typical exercise from the textbook for key objectives Brad has been exposed to. Have one or two exercises for each objective. If he can do these, go to the next objective. If not, don't assign him other exercises of the same type. For example, if he cannot do two-digit multiplication, such as 26×37, don't ask him to determine 35% of 68.

From these questions, teacher conferences, and the checking of Brad's work, you and the teacher can identify objectives he has mastered as well as those he has not mastered. It is helpful to list these objectives for easy reference and to keep this list in a folder along with the student's work and a tutoring log.

THE TUTORING LOG

The tutoring log should contain a list of objectives for the session, a record of objectives accomplished, and a summary of the activities of the session, including difficulties encountered and the assignment given. Keeping this information in chart form is an effective aid for organization and makes easy reference possible.

An example of a log entry for a session with a first-year algebra student is shown in figure 1.

Date	Objective(s) of Session	Objective(s) Accomplished	Activities of Session	Assignment Given
12/5	To develop skill in adding like terms using the distributive axiom	Yes	1) Used concrete examples: 5 dollars plus 7 dollars 2) Reviewed distributive axiom: $(a + b)c = ac + bc$ 3) Practiced adding: $5x + 7x$, $-7x + 3x$, $x + 15x$	p. 21 (3, 7, 12, 15)

Fig. 1

An example of consecutive sessions for another first-year algebra student is given in figure 2. Note that follow-up activities from the previous session are included.

DIAGNOSIS OF STUDENT DIFFICULTIES

Mathematics is a highly structured subject. Often students cannot perform a certain operation because they haven't achieved the skills needed

Date	Objective(s) of Session	Objective(s) Accomplished	Activities of Session	Assignments
10/12	1) To identify expressions of the form $a^2 - b^2$	Yes	1) Reviewed proof of theorem $a^2 - b^2 = (a - b)(a + b)$	*Algebra I* p. 36 (12–18, 20, 22)
	2) To develop skills in factoring this type of expression: $a^2 - b^2 = (a - b)(a + b)$	Yes	2) Used concrete examples to illustrate theorem	
			3) Practiced writing perfect squares in b^2 form; example: $36 = 6^2$ $a^4 = (a^2)^2$	
			4) Practiced factoring expressions of the form $a^2 - b^2$; examples: $49 - x^2$ $4x^2 - 16$	
10/14	1) To identify expressions of the form $a^3 - b^3$	Yes	1) Reviewed objectives of session on 10/12 by practicing examples	p. 38 (18–20, 24, 28)
	2) To develop skill in factoring expressions of the form $a^3 - b^3 = (a - b)(a^2 + ab + b^2)$	No (time ran out)	2) Checked homework assignments and helped Al correct mistakes	
			3) Introduced factoring of difference of cubes. Had Al identify expressions that fit the form $a^3 - b^3$	
			4) Al could not identify the "*a* and *b*" in the difference of cube type of expressions. Therefore, we practiced placing perfect cubes in a^3 form	

Fig. 2

in that operation. For example, pupils probably will not master division until they become proficient with subtraction and multiplication. Algebra students won't be proficient at adding polynomial expressions until they can add, subtract, and multiply integers.

When the key objectives that Brad has not mastered are identified, you should check his ability to perform the operations necessary for each objective. For example, if he cannot add two-digit numbers involving regrouping (56 + 37), check to see if he knows the background material. To succeed in this objective, he should know basic addition facts (6 + 7 = 13), use place-value concepts (tens, ones), and do regrouping (6 + 7 = 1 ten and 3 ones).

When a student fails to complete a computation or follow a procedure correctly, do not teach the entire process over again. Instead, find out what part, or parts, of the process are causing trouble and reteach that part.

EXAMPLES OF DIAGNOSES OF STUDENT ERRORS

Tom may continually say 7 × 8 = 54. Some observers would conclude that he doesn't know the basic multiplication facts, but this is probably false. Tom may know all but a few of the 100 basic facts (0 × 0, 0 × 1, . . . , 1 × 0, 1 × 1, . . . , 9 × 9). Spend time only on the few facts that are giving Tom difficulty.

When a student has trouble with skills like long division or multiplication, don't start by reteaching the entire skill. Watch the student do the process and look for specific errors.

Sue's work for the exercise 86 × 45 is shown in figure 3. Sue has the wrong answer, but don't conclude that she can't multiply! Examine her work carefully to find her error.

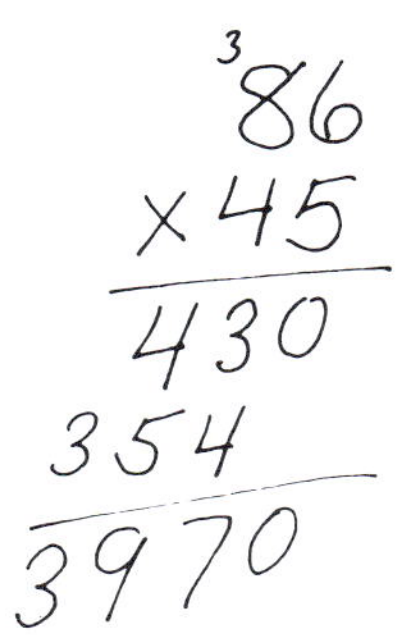

Fig. 3

Sue used the "carry" digit, 3, for both partial products. Every other step is correct. Sue needs help with only one step.

Lynn recorded this result: 4896 ÷ 32 = 1413. Apparently Lynn can't perform the division process. Let's see what she did (fig. 4).

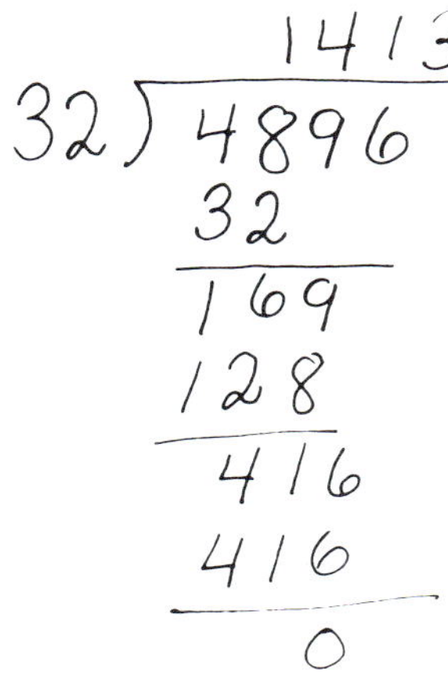

Fig. 4

Lynn failed to notice that 169 — 128 = 41 is greater than the divisor, 32. Her error occurred in working with 416. Lynn found that 32 × 13 = 416, but the 13 does not belong in the location shown in figure 4. One error is all that needs to be worked on.

Everyone makes computational errors from time to time, but as a tutor, you must watch for any error *consistently* made by your student.

Often the mastery of a given objective is based on intense practice of prerequisite skills. Seek assistance from teachers, textbooks, and professional references to determine the prerequisite skills and concepts needed for a particular objective.

Preparation for Tutoring Sessions

IN ORDER to effectively and efficiently help someone learn, a tutor must plan for each session. This planning should include the following steps:

1. An objective for the session should be clearly stated.
2. A method of presentation, along with possible alternatives, should be identified.
3. Exercises should be selected for the student to use as practice.
4. Evaluation, related to the objective, should be made.

In mathematics, the textbook often includes these steps in its organization. Each section is based on one or more objectives. A presentation is suggested by the format of the section. A set of exercises is available for practice. Often these parts are spelled out clearly in the teacher's manual. Once an objective is identified, a teacher will frequently add alternative presentations to those suggested, make selections from the available exercises, or add additional ones for student practice.

As the tutor, you should employ the same basic planning strategies.

IDENTIFY THE OBJECTIVE

Tutoring sessions are not isolated events. Each session must be based on the progress made at previous sessions and on the goals for the tutoring program. Recall the student's progress at previous sessions. Check his or her classwork. Review the goals for tutoring. Refer to the folder containing objectives, student work, and the tutoring log. With this background you should identify specific objectives for each session and discuss them with the student. The objectives should be stated so that success can be reached in a short period of time and this success can be identified by the student during that session if at all possible. For a typical session, one or two objectives are sufficient. A plan designed to achieve more objectives is unreasonable and can cause confusion and frustration for both of you.

IDENTIFY ALTERNATIVE METHODS OF PRESENTATION

After an objective is clearly identified, you should seek a variety of ways to meet it. First, examine the text's presentation. Then seek alternative ideas from mathematics teachers, from other texts in the library or mathematics resource center, or from other tutors.

Your effectiveness as a tutor depends on your ability to present a topic in different ways. Although the presentation in the text may be clear to you, most likely it was unclear to your student. Try to discuss the presentation in your own words. Use words, ideas, and examples acceptable in your age group; don't stick to formal mathematical terminology.

When you identify an alternative approach to your objective, examine the approach for consistency with your previous work. To use the new idea, you must make it fit your student's previous understandings.

In addition, your planning should take your student's interests into consideration. For example, if the student likes football and the objective of the tutoring session is to add positive and negative integers, it would be appropriate to describe a football game where losses and gains can be represented as integers. Appealing directly to his or her interest could well spark enough enthusiasm to meet the objective quite quickly.

Including applications related to the objective is also a technique that may create interest and enthusiasm. Again, if the objective is to add integers, relating it to temperature changes reported on the local TV weather forecast will demonstrate an immediate use of integers.

All of us appreciate musicians or athletes who can alter their performance to meet a variety of demands. Don't forget that they worked endless hours preparing alternative routines for such conditions. The successful tutor or teacher has prepared in this way too.

PLANNING THE SESSION

A tutoring session will be more valuable to the student if you have an initial plan than if you have no sequence of activities in mind. The use of time is crucial to all learning activities: classroom instruction, athletic practice, music lessons, and your tutoring sessions. If you must deviate from the plan, you will still be more successful than if you are operating with no plan.

A SHORT REVIEW

The first few minutes of the session are best used to conduct a short review and to develop a positive attitude in your student. After a friendly greeting, start a short review on work the student has successfully accomplished. This gets the student started on a positive note, and you can then move into a check on his or her current progress. This can be done by examining work the student did between sessions or by having your student do several exercises on the spot.

The early work should be selected so that the probability of success is high. This will help the student develop self-confidence and provide motivation for working throughout the period.

The review is often necessary, especially for those students having difficulties, to order their thoughts and patterns for mathematical reasoning. This review can also be used to prepare them for the next objective. For example, when the objective is to multiply positive and negative integers, a review on addition is a valuable introduction.

THE LESSON

The instruction you plan should be brief, to the point, and aimed at the student's level of understanding. If your student is typical, his or her attention span will be limited. Your explanations should seldom last more than a few minutes; don't expect your student to sit passively through a long, involved explanation. Present the topic clearly, through examples whenever possible. Then plan to expand the topic based on your student's questions and his or her ability to perform the work you have planned.

An activity or discovery lesson may last longer. This lesson actively involves the student in doing examples, trying experiments, or answering questions. This approach will give you the opportunity to gain continuing feedback from the student.

TIME TO WORK

After introducing a topic, you should provide exercises for the student to do during the session. For example, if Jon says he knows how to do the work, let him try it. His performance is a most important phase of the session. If he is successful, you can tell him so immediately. This positive feedback is very important to someone having difficulties.

If your pupil is having trouble, you can diagnose the difficulty while there is time to correct the situation. At times, you will find it necessary to reteach a topic with an alternative explanation. More frequently, however, you may just need to help with a small part of the procedure. Here a diagnosis of student work is important. For example, Sue, who is mastering long division, may be able to do everything correctly until she obtains the last remainder. In this situation, don't reteach the entire process. Just review the final steps with her.

PRACTICE WITH A PURPOSE

The practice of mathematical processes is a crucial phase for most of us. It is especially important to the student who is having difficulty or lacks confidence. When you have presented a topic and your student has shown an initial understanding of it, have him or her practice it.

Practice is most effective when students understand what they are practicing. To practice without such an understanding will produce little but rote memorization and poor retention.

Most frequently, practice takes the form of a dittoed sheet or a set of exercises in the text. When a more motivating type of practice is possible, it is to be favored. A game like Battleship or an activity of challenge such as Contig can provide exciting practice for a specific objective. Sources of many games and activities appropriate for mathematical practice are listed in the Bibliography.

Following this practice, a set of exercises should be assigned for the next tutoring session. Make the purpose clear to the student: the assignment will provide an opportunity to refine the skill learned during the session. A word of warning must be made here. Be careful to select exercises that are directly related to the objective of the tutoring session and that the student can do successfully. Assigning exercises 2 through 50 on page 396 may be inappropriate practice.

At the conclusion of the session, careful notes should be entered in your log concerning the objectives reached during the session, difficulties encountered, and the assignment given. From this log the next session can be planned.

REVIEW FOR THE TUTOR

To be an effective tutor, you must have prior to each tutoring session an understanding of the objectives you are trying to reach. If you are asked to teach a concept with which you are not as familiar as you should be, a review prior to planning the session is most important. Do not hesitate to ask for help from a teacher or another tutor. It will be an opportunity to grow mathematically at the same time you are preparing for the session. Valuable tutoring time should be spent assisting the student, not learning what you are to teach.

The Tutoring Session

IN PREVIOUS sections, suggestions for establishing rapport with the student, identifying objectives for instruction, locating a tutoring station, and planning for a session have been provided. This chapter will build on these ideas and provide specific guidelines for conducting effective tutoring sessions. The two most important qualities tutors must develop are (1) a good working relationship with their students and (2) the capability to help them reach their goals. Seldom can one happen without the other. The following guidelines are addressed to these factors.

BE PROMPT AND DEPENDABLE

You can set a tone for work and success by being on time and ready for business. If you cannot attend a planned session, notify Brad as early as possible. Your dependibility affirms your concern about his progress.

START WITH A REVIEW

An important diagnostic procedure is to determine at the start of the session what Brad can do. Typically, this involves checking work assigned in the previous session or in class. It is advisable to follow this with an opportunity for him to do an exercise similar to one found in the homework. He may have done the work one or more days earlier. See if he can still do it. This will indicate to you his current status, and it will be an excellent review for him.

SET OBJECTIVES FOR THE SESSION

Clearly state one or two specific objectives for the session based on the goals for the tutoring experience and Brad's performance. Be sure these objectives are realistic. Brad will be measuring himself on the basis of your expectations. If you expect too much, he may feel defeated. If he can see that he has mastered an objective, he will develop the self-confidence necessary to attempt the next one.

BE FLEXIBLE IN THE SESSION

With a well-developed plan, you should be able to modify activities and

expectations; don't forge ahead just to keep on your teaching schedule. If Brad is not progressing as fast as you had planned, change your activities to assist him.

At times, he may become greatly discouraged or frustrated during a tutoring session. Back off; don't lose your cool. You are well advised to set aside your plans and talk over the matter. When you return to tutoring later in the session, start with activities that are likely to provide success.

USE PRIOR SUCCESSES TO CREATE INTEREST IN NEW MATERIAL

Mathematics is a highly structured discipline. The sequenced content is the cause of the difficulties that many individuals encounter. A student who fails to understand today's work will typically fail at tomorrow's assignment too. But frequently you can use the sequence to advantage in the tutoring session.

If you have carefully analyzed the skills necessary for a given objective, you can start with those that Brad already knows. Have him do several of these exercises. Then show him how you are advancing just one step. For example, to begin instruction on the addition of fractions with different denominators, be sure he can add fractions with the same denominator. Then indicate that the same procedure is followed once the two fractions are rewritten with common denominators. This strategy is important throughout mathematics instruction. Use it whenever possible.

INVOLVE THE STUDENT IN DISCOVERING NEW CONCEPTS THROUGH EFFECTIVE QUESTIONING TECHNIQUES

When you seek to teach Brad a mathematical concept (a mental abstraction), you must get him involved in the learning process. It is not advantageous to lecture to him about the addition of positive and negative integers. It is more effective to build up this idea through examples based on his experiences.

Pose questions that he can use to develop the concept. Ask, for example,

"If you take five steps north and three steps south, how far are you from the starting point?" Such questions can help him develop a concept for himself. After several similar questions, he may be able to generalize a way to add positive and negative integers—in his own words.

In developing questions for the discovery of a concept, avoid those that can be answered with a simple yes or no. Often these are poor questions that do not encourage thinking by the student. Besides, he has a fifty-fifty chance of getting them right without understanding the concept. Develop questions that encourage the student to demonstrate his understanding of a concept.

If your question requires some thought, don't expect an answer instantantly. Be patient. Give your student adequate time to absorb your question, think it through, and formulate a verbal response. This takes time!

When Brad responds, listen carefully. He is telling you a lot about his understanding of the concept. Don't be thinking of your next question while he is answering; it is too easy to miss his answer. To base your next question on his answer demands that you be a good listener.

Avoid questions like "Do you understand?" or "Do you have any questions?" It is hard for a student to answer honestly that kind of question. Give him an example to work. He will show you and himself whether he understands the idea.

GIVE A BRIEF EXPLANATION OF THE NEW PROCESS; THEN GET THE STUDENT INVOLVED

When you try to teach someone how to do a mathematical procedure, make the explanation and examples short and to the point. Don't go on and on and on. If you are teaching a formal procedure for division, explain it clearly through one or two well-chosen examples. Now stop and let Brad try a carefully selected exercise. With an exercise that is too complex and includes ideas that have not been previously taught, he can become frustrated. With one that is too simple, he can answer it without really understanding the process.

It is important that your student get involved without great delay. After your brief, clear explanation, Brad can show you how much he picked up by trying a simple example. His work will provide feedback for you to use in choosing more complex exercises or in reteaching parts of the original process.

A serious flaw in most teaching is too much talking. It is easy to present more than the student can grasp. Stop! Let him show you and himself if he is with you.

Base your next move on Brad's work. If he can do half of the process you are presenting, use that success to concentrate on the points of misunderstanding. Don't reteach the entire process.

LET THE STUDENT KNOW HOW HE IS DOING

During the tutoring session, it is important to let Brad know if he is reaching the objective that has been set. He can often get a good idea of his progress from his ability to answer your questions and to do the exercises you provide. In addition, you should acknowledge his successes. Don't artificially praise him on every right answer, but let him know you are pleased with his ability to do an exercise that demonstrates the mastery of an objective.

If Brad is not picking up a concept as fast as you had planned, you may wish to report that fact in a positive way: "This is a complicated concept. It takes time, but we'll get it." You have factually reported his progress, and you have indicated that you haven't given up.

CHECK MATHEMATICS WORK: DON'T DO IT

Some time in a tutoring session may be scheduled for checking work Brad has attempted for class or the session. It is often appropriate to examine this work carefully. As the tutor, you must resist doing the exercises. Help the student identify errors or faulty reasoning in the work.

A valuable guide for the tutor is to refrain from picking up a pencil during this correction period. By picking up the pencil, you are indicating that Brad is unable to do it himself. Help identify errors; question your pupil on how he might revise his work. Then let him do the revision.

PROVIDE AN ASSIGNMENT FOR THE NEXT SESSION

Before the session ends, schedule the next one. Clearly identify the date, time, and place of your next meeting. To increase the value of tutoring sessions, it is advisable to have Brad do some work between sessions. Sometimes this will be a specific assignment for the session, regular classwork, or preparation for an examination. Help him understand the purpose of the assignment. The importance of an assignment should be reinforced by following up on it in the next session.

END THE SESSION ON A POSITIVE NOTE

No matter what has happened during the session, you should seek to end it with success for the student. He can often live with limited frustration and confusion in the period if he feels confident as he leaves.

A frequent practice is to provide an assignment on the objectives for that day. Give the assignment with adequate time for Brad to get a good start on it. This allows time for reteaching if necessary. Strive to have the session end with the student completing several exercises correctly.

Evaluation

EVALUATION DURING SESSIONS

As preceding chapters have implied, evaluation is a continous process during each of the tutoring sessions. Evaluation should occur in several ways and for several purposes. The purposes for evaluating the student during the tutoring session include the following:

1. To aid the formulation of the next question
2. To determine the selection of the next example or problem
3. To determine whether additional work on a specific objective is necessary
4. To determine whether the reteaching of an objective is necessary
5. To determine whether alternative approaches should be introduced to aid in reaching an objective
6. To determine whether a more complex objective may now be introduced
7. To demonstrate to the student his progress and the worth of the tutoring sessions

Evaluation in its most positive form can be used for motivating, reinforcing, and building confidence.

An evaluation program should be developed jointly by the mathematics instructor and the tutor. Typically the teacher will want to give a test when certain objectives have been mastered in the tutoring sessions. As the tutor, you should periodically evaluate the student's skills in a testlike setting. On the basis of this evaluation, you can more accurately report your student's progress to the instructor. This report will permit the teacher to give the appropriate test at the correct time. After the test, you should ask the teacher to interpret the results in terms of the objectives you have taught. This discussion should give you direction on how to plan future sessions. The teacher may want you to place more emphasis on certain objectives or reteach others. Results of these evaluations should be communicated to the student and, when appropriate, to the parents.

EVALUATION BY THE TUTOR

You, the tutor, should provide meaningful evaluation throughout the time over which the tutoring is taking place. Your evaluation can take several forms:

1. An oral check on definitions and procedures
2. A written test based on the objectives of an individual tutoring session
3. An observation of changes in the student's attitude

Putting your student under a simulated test situation can be a worthwhile activity, but constructing written tests is not easy. Carefully study sample tests in the student's textbook. Examine previous tests given by the student's mathematics instructor and try to follow a similar format. Select problems for your test only after you have worked them yourself. As you are working the problems, check to see that they test the objectives you intended them to test. Ask yourself these questions: (1) Is the problem fairly representative of the level of difficulty the objective implies? (2) Does the solution of the problem demand skills that have not been introduced to my student? If the answer to the second question is yes, revise your question.

Determine, with the help of the instructor, the appropriate length of time in which your student is to complete the evaluation. The importance of speed on a test should be discussed with the teacher. The time you allot for the test should be similar to that allowed by the teacher if the testing situation is to be a true simulation.

Observing changes in student attitude is a more difficult form of evaluation often overlooked by the tutor. Are there positive changes in attitude toward learning the mathematics required? Is the homework being completed? Is the student on time for class? Does the student consistently attend class? Does the student show increased concern about the problems that he or she has done incorrectly? Positive changes in attitude may indicate that the tutoring experience has been successful in giving the student confidence in working with a greater share of the course objectives.

Sharing the results of your evaluation with your student's mathematics teacher will be beneficial. The teacher can help interpret these results and again give you direction in planning future sessions. Moreover, the teacher will gain some additional insight into the student's progress.

In some situations, the evaluation may point out a lack of success in the tutoring program. If this happens, a discussion with the teacher should be held on whether the tutoring sessions should continue and whether or not you are the appropriate tutor for the student. Tutoring is by no means 100 percent effective with all students in all mathematics courses. As in any human interaction, not all tutors are effective with all students.

With reasonable effort on the part of tutor and student, a tutoring program will yield valuable benefits to all concerned. Frequently, the major goals set by you and the student will be achieved. If major goals are not reached, you should reevaluate the tutoring experience in terms of what intermediate goals your student *has* mastered. It is important for the student—and you—to see the growth he or she has made in mathematical competence.

SUMMARY CHECKLIST

Planning and Conducting the Early Sessions

1. Arrange a conference with the student's instructor to determine objectives for the student to master.
2. Approach the sessions with confidence.
3. Develop a positive atmosphere.
4. Determine the reason why the student sought a tutor.
5. Acquaint yourself with the student's interests.

Planning for the Session

1. Determine objectives for the session based on the student's progress and the instructor's guidelines. Refer to your tutoring log.
2. Review the objectives you plan to teach.
3. Consult with the instructor to locate resource materials and supplementary exercises.
4. Construct a review activity for the purpose of checking on objectives that have been previously taught.
5. Construct an instructional lesson for the objective(s) based on suggestions from the teacher and resource materials.
6. Construct an appropriate, well-selected set of exercises for the student to complete prior to the next tutoring session.
7. Arrange for a tutoring location—one without distractions.

Conducting the Session

1. Be prompt.
2. Develop a positive atmosphere.
3. Sit beside the student.
4. Start with a review of objectives previously taught.
5. Discuss the objective(s) for the session.
6. Explain briefly the new process to be introduced.
7. Involve the student as soon as possible.

8. In checking your student's work, let the student make the corrections with your guidance. Avoid picking up the pencil unnecessarily.
9. Keep your student informed about his progress during the session.
10. Assign an appropriate, well-selected set of exercises for the student to complete prior to the next tutoring session.
11. End the session on a positive note—a successful experience.

Follow-up to the Session

1. Enter information on your student's progress and learning difficulties in your tutoring log.
2. Report progress and evaluation to the student's instructor.

Preparations for Tests and Examinations

HELPING your student prepare for tests is often a task assigned to you, the tutor. Here is a brief list of study habits and techniques for students to use at test preparation time. This list is not complete, and you may want to add your own techniques to it.

1. Prepare a list of specific objectives to be included on the test.
2. Locate and work examples and problems related to each of the objectives you have listed.
3. Learn definitions of new words. Being able to repeat these definitions orally is a good check to see if you really can recall the appropriate definition. When learning a definition, note its specific limitations.
4. Begin a review of the objectives several days prior to the test, tackling a few of them each day.
5. Look for and try to recall general patterns. Study previous quizzes and, in particular, problems that you did incorrectly.
6. Select problems from past homework assignments. Rework these problems. Compare the results with your original homework results.
7. Don't just look over your homework or previous tests. It is much more important that you pick up a pencil and do the problems again.

Summary

IN DEVELOPING an effective learning setting for your student, review these guidelines, alter them to meet your situation, and try your own techniques. Successful tutoring, like teaching, is not an exact science: it is based on thorough planning and good communication between student and tutor. With continued effort by both, little daily frustrations are usually overshadowed by the student's noticeable growth in mathematical competencies.

Annotated Bibliography

Mathematics methods

Arithmetic Teacher. Reston, Va.: National Council of Teachers of Mathematics. Published monthly except June through September.

A valuable source of teaching techniques and instructional suggestions for many topics at the elementary school level.

Ashlock, Robert B. *Error Patterns in Computation: A Semi-Programmed Approach.* Columbus, Ohio: Charles E. Merrill Publishing Co., 1972.

A good introduction to identifying student errors in mathematical computations.

Biggs, Edith, and J. R. MacLean. *Freedom to Learn.* Don Mills, Ont.: Addison-Wesley (Canada), 1969.

An excellent presentation on student learning through involvement in concrete mathematical situations.

Butler, Charles H., F. Lynwood Wren, and J. Houston Banks. *The Teaching of Secondary Mathematics.* 5th ed. New York: McGraw-Hill Book Co., 1970.

A thorough examination of instruction for mathematical topics.

Dumas, Enoch. *Math Activities for Child Involvement.* Boston: Allyn & Bacon, 1971.

Excellent teaching techniques for whole numbers, fractions, and decimal instruction.

Fawcett, Harold P., and K. B. Cummins. *The Teaching of Mathematics from Counting to Calculus.* Columbus, Ohio: Charles E. Merrill Publishing Co., 1972.

Strong presentation of teaching techniques for mathematical topics.

Johnson, Donovan A., and Gerald R. Rising. *Guidelines for Teaching Mathematics.* 2d ed. Belmont, Calif.: Wadsworth Publishing Co., 1972.

A superior book of mathematical methods and suggestions for specific topics.

Kidd, Kenneth, S. S. Myers, and D. W. Cillay. *The Laboratory Approach to Mathematics.* Chicago: Science Research Associates, 1970.

A resource of activities for student involvement in learning mathematics and attitude improvement.

Mathematics Teacher. Reston, Va.: National Council of Teachers of Mathematics. Published monthly except June through August.

A valuable source of suggestions on mathematical topics and teaching techniques at the secondary school level.

National Council of Teachers of Mathematics. *Games and Puzzles for Elementary and Middle School Mathematics: Readings from the "Arithmetic Teacher."* Edited by Seaton E. Smith, Jr., and Carl A. Backman. Reston, Va.: The Council, 1975.

A variety of games and puzzles intended to provide nontraditional ways to practice key mathematical processes.

_____. *The Growth of Mathematical Ideas, Grades K–12.* Twenty-fourth Yearbook. Washington, D.C.: The Council, 1959.

An excellent presentation of concept development through grade levels.

_____. *The Slow Learner in Mathematics.* Thirty-fifth Yearbook. Washington, D.C.: The Council, 1972.

Specific suggestions for working with students having difficulties in mathematics. Methodology and content topics are well covered.

_____. *Teacher-made Aids for Elementary School Mathematics: Readings from the "Arithmetic Teacher."* Edited by Seaton E. Smith, Jr., and Carl A. Backman. Reston, Va.: The Council, 1974.

A variety of suggestions for the construction and use of manipulative materials to help individuals develop basic mathematical concepts.

_____. *The Teaching of Secondary School Mathematics.* Thirty-third Yearbook. Washington, D.C.: The Council, 1971.

Excellent chapters on motivation, teaching concepts, skills, proof, and problem solving. Practical examples are included.

Paige, Donald D., R. E. Willcutt, and J. M. Wagenblast. *Mathematics Methods for Junior High School Teachers.* Boston: Prindle, Weber & Schmidt, 1968.

A source of alternative instructional techniques.

Reisman, Fredericka K. *A Guide to the Diagnostic Teachings of Arithmetic.* Columbus, Ohio: Charles E. Merrill Publishing Co., 1972.

Practical suggestions for diagnosing student problems in arithmetic.

Reys, Robert E., and T. R. Post. *The Mathematics Laboratory: Theory to Practice.* Boston: Prindle, Weber & Schmidt, 1973.

Excellent examples of laboratory activities in mathematics. Valuable sequencing techniques for mathematics learning.

Sawyer, W. W. *The Search for Pattern.* Baltimore: Penguin Books, 1970.

Numerous examples of activities for student discoveries.

Schminke, C. W., Norbert Maertens, and William R. Arnold. *Teaching the Child Mathematics.* Hinsdale, Ill.: Dryden Press, 1973.

One of many excellent resources for content and teaching strategies for mathematics work up to algebra.

Westcott, Alvin M., and J. A. Smith. *Creative Teaching of Mathematics in the Elementary School.* Boston: Allyn & Bacon, 1967.

Alternative teaching strategies and computational procedures for basic arithmetic instruction.

Content materials

Alberta Teachers' Association, Mathematical Council. *Active Learning in Mathematics: A Set of Resource Materials for Teachers.* 1970 Annual, edited by D. B. Harrison, Edmonton, Alta.: The Association, 1970.

Activity materials for instruction.

Association of Teachers of Mathematics. *Notes on Mathematics in Primary Schools.* London: Cambridge University Press, 1967.

Examples of mathematics topics appropriate for junior high school programs in the U.S.

Jacobs, Harold R. *Mathematics: A Human Endeavor.* San Francisco: W. H. Freeman & Co., 1970.

A valuable source of innovative and creative teaching strategies.

Johnson, David C., Dale E. La Frenz, Larry L. Hatfield, Thomas E. Kieren, John W. Walther, and Pamela W. Katzman. *Computer Assisted Mathematics Program.* Glenview, Ill.: Scott, Foresman & Co., 1968.

Excellent source of key mathematical concepts developed with the use of elementary computer programming and flowcharting.

Johnson, Donovan A., Viggo P. Hansen, Wayne H. Peterson, Jesse A. Rudnick, Ray Cleveland, and L. Carey Bolster. Activities in Mathematics series: First Course, Second Course; Applications in Mathematics: Course A, Course B. Glenview, Ill.: Scott, Foresman & Co., 1972.

Excellent activities and worksheets for student use.

Krulik, Stephan. *A Handbook of Aids for Teaching Junior-Senior High School Mathematics.* Philadelphia: W. B. Saunders Co., 1971.

Suggested aids and models for developing mathematical topics.

National Council of Teachers of Mathematics. *Topics in Mathematics for Elementary School Teachers.* Twenty-ninth Yearbook. Washington, D.C.: The Council, 1964.

______. *More Topics in Mathematics for Elementary School Teachers.* Thirtieth Yearbook. Washington, D.C.: The Council, 1968.

Thorough sources on mathematics content in the elementary and junior high school programs.

Schaum's Outline Series Including Theory and Solved Problems. New York: Schaum Publishing Co., n.d.

The various mathematics titles in this series include hundreds of completely solved and explained problems. Helpful for student practice.

Seymour, Dale. *Aftermath.* Vols. 1–4. Palo Alto, Calif.: Creative Publications, 1970.

A series of enrichment books designed to reinforce the regular curriculum, including puzzles, games, and creative drill sheets.

Sobel, Max A., and E. M. Maletsky. *Teaching Mathematics: A Sourcebook of Aids, Activities, and Strategies.* Englewood Cliffs, N.J.: Prentice-Hall, 1975.

Superior source of ideas for motivational instruction. An extensive set of interesting mathematical topics for use by tutors.

General tutoring

Dadourian, H. M. *How to Study How to Solve.* Reading, Mass.: Addison-Wesley Publishing Co., 1957.

A service type of booklet for secondary and college students designed to be used during the initial sessions of a mathematics course. A basis for discussion of effective ways of studying mathematics subjects and solving mathematics problems.

Lippitt, Peggy. *Students Teach Students.* Fastback 65. Bloomington, Ind.: Phi Delta Kappa Educational Foundation, 1975.

Margenau, James, and Michael Sentlowitz. *How to Study Mathematics.* Reston, Va.: National Council of Teachers of Mathematics, 1977.

A booklet concerned with the problem of how to study mathematics in junior high school through high school. Emphasis is on the areas of homework, classwork, and tests.

Rauch, Sidney J., ed. *Handbook for the Volunteer Tutor.* Newark, Del.: International Reading Association, 1969.

A thorough presentation of tutoring competencies and suggestions; stress is on reading situations.

Mathematics textbooks

The most accessible source of objective-related material will be mathematics texts in your school. Texts collected by the staff and book-selection committees from prior years provide excellent alternative presentations and well-chosen examples. Check the faculty's professional library and individual teachers' collections.